SHARK ATTACK Britain

Shark Cornwall Publishing

Published by:
Shark Cornwall, Dulverton House, Crooklets,
Bude, Cornwall, EX23 8NE.

First published October 2010

ISBN no. 978-0-9558694-2-6

Photographs: Shark Bay Films
Photo enhancement: Mike Peirce
Design and Artwork: Anthony Moulton
Printed and bound in England by
SR Print Management Limited, Watling Court, Orbital Plaza,
Cannock, Staffordshire. WS11 0EL

– – – – – –

SOON TO BE PUBLISHED BY THE SAME AUTHOR
"Shark adventures – the expeditions"
"Pocket guide to Sharks in British Seas"
"Sharks in British Seas" (second edition)

OTHER PRODUCTS
"Sharks in British Seas" - Book (Richard Peirce)
"Sharks in British Seas" - DVD (Simon Spear - Richard Peirce)
"Sharks off Cornwall and Devon" - Book (Tor Mark Publishing)
Pirates of Devon & Cornwall - Book (Richard Peirce)
All books and DVD's can be bought through our website
www.sharkconsoc.com

SHARK ATTACK BRITAIN

When researching the book and film versions of *Sharks in British Seas* I started coming across 'shark attack' incidents which ranged from the bizarre to the dangerous, the ridiculous, and the tragic.

In 2009 I got together with John Boyle of Shark Bay Films to assess the viability of making a film about British shark attacks. We wanted to make something that was different and look at sharks and the subject of shark attacks in ways that hadn't been done before. Our task was made easier by having some amazing stories to tell and by the fact that many Britons are unaware there are sharks in UK waters, let alone that there have been shark attacks. We explored the realities of who attacks who, who eats who, who is guilty and who is innocent?

This short book follows the film *Shark Attack Britain* and tells all the same stories and a few more. All the photos used are frame grabs from the film, and like the film the book sets out to entertain, inform and shock.

This book is dedicated to all those working
in shark conservation, trying to make a difference
and ensure a future for this threatened group
of animals.

- - - - - -

Most of the illustrations in this book are frame
grabs from the film *Shark Attack Britain*.

CONTENTS

Chapter 1

DEATH BY DROWNING

Growing to 10 metres in length and weighing the same as a single decker bus, the plankton-eating Basing shark is generally thought to be harmless. It is the second largest fish in the ocean and now has protection in British waters, which appears to be allowing numbers to recover.

It has always seemed ironic to me that in the early 21st century we know virtually nothing about an enormous creature living on our own planet, and yet we put men into space in the 1960's. Over fifty years since men walked on the moon and we have only now discovered that British Basking sharks don't hibernate on the ocean floor during the winter, and that they make trans-Atlantic migrations!

The first human deaths caused by sharks in British waters, occurred off Carradale in western Scotland in the Kilbrannon Sound over seventy years ago, in September 1937.

Basking sharks were much more plentiful than they are now; indeed Basking shark hunting only stopped in western Scotland in the 1950's. In those days if you looked across from Carradale to the peaks of Arran you would regularly see large numbers of Basking sharks fins.

I first came across the 'Carradale' incident when I was researching my book *Sharks in British Seas*. The book was followed by the film of the same name, and we decided to go there to investigate and see if there were any witnesses or even survivors still alive.

"Chrystal and Archie Patterson being interviewed."

My visit to Carradale resulted in the footage used in *Sharks in British Seas*. We heard the story at first hand but on that visit didn't have time to interview Chrystal Patterson who is the closest living relative of one of the survivors and of two of those who died. I returned later to interview her for *Shark Attack Britain*.

Next to Carradale is Port Righ which is a small sheltered inlet facing east. Chrystal Patterson and her husband Archie, a one time Basking shark harpoon gunner, live between Carradale and Port Righ.

Captain Angus Brown, Chrystal's uncle, was originally from Carradale but had moved to Swansea. Captain Brown and his family used to return to Carradale for their summer holidays. His brothers Robert and Archibald had stayed in Carradale and were fishermen. Their largest boat was the 15 foot *Eagle* which could either be rowed or sailed. Donald MacDonald was a 14 year old local boy who worked for Robert Brown during the holidays.

There had been a few days of stormy weather, but on Wednesday September 1st the weather and sea conditions improved. It was overcast with sunny spells, there was only a little wind but there were still swells from the south. There were a lot of Basking sharks in the Sound and many of them were seen breaching (jumping

"The shark breached underneath the boat and overturned it".

clear of the water). Angus and Robert Brown took the children, Jessica and Neil, out sailing in the *Eagle*, and young Donald MacDonald went with them. Angus, Jessica and Neil were sitting in the stern, Donald MacDonald was amidships with the oars, and Robert Brown was in the bows.

The little *Eagle* had her sail up and was making good speed out of the bay when the main halyard snapped and the sail came down. The two men took the oars and started rowing back towards Port Righ.

Suddenly the boat seemed to lift out of the water, there was a splash and then the *Eagle* was on her side with her mast at an acute angle. Donald MacDonald said that he found himself 9 or 10 feet under the water with the others. Robert and Angus righted the boat and put Jessica and Neil into it. Then the shark which had capsized the boat came back and did it again – there are accounts which say the boat was capsized three times. Jessica and Donald clung to the hull and tried to drag Angus onto it, but he died while they were holding onto him.

The incident had been seen by many ashore and several boats sped to the rescue. Jessica and Donald survived but Robert, Angus and Neil all drowned, Robert's body not being recovered for several days. Newspaper reports suggested the boat had been holed and the attacks were deliberate. Donald MacDonald remembers the hull was 'bruised and marked' but intact. There is no doubt that the *Eagle* was capsized by a Basking shark breaching underneath her, equally there is little doubt that this was a tragic accident and not a deliberate attack.

Basking sharks are thought to breach as part of mating displays and to rid themselves of parasites. Sharks had been seen breaching all morning on that fateful day. Should an animal as large as a Basking shark decide to charge and attack a little boat like the *Eagle* it would cause very considerable damage. That the *Eagle* appears to have only been struck a glancing blow further supports its having been an accident.

At the time of the incident the *Eagle* was in 26 metres of water which is deep enough for a shark to breach. Donald MacDonald remembers that the boat seemed to stand right up on her end, which suggests a strike from underneath by a breaching shark.

Another theory is that one of the oars may have touched the shark and caused a tail flick which overturned the boat. It's possible but at odds with Donald's 'standing on it's end' statement. We will never know exactly what happened but the best guess has to be that the first capsize was caused by a breaching shark, and the subsequent overturnings may have been tail flicks from the stunned and disorientated animal.

One of the rescuers reported that when he arrived on the scene the shark was still at the surface going round the boat lashing its tail. This further indicates that the animal may have been injured by its encounter with the *Eagle*. Chrystal Patterson is the cousin of Jessica Brown who survived. In her interview for the film *Shark Attack Britain* she describes the incident as follows.

"There were five people in the boat, my uncle Robert, Uncle Angi, my cousin Jessica, my wee cousin Neil, and local man Donald McDonald. They went out fishing which was the big treat when they came on holiday. They were just off the point at the end of the bay and the shark came at them. It jumped below the boat and upset the boat. Uncle Angi and Uncle Robert put wee Neil back in the boat, and Jessica and Donald McDonald. But the shark came back a second time and upset the boat again, they weren't so lucky that time. Uncle Angi was drowned, Uncle Robert was drowned and wee Neil was drowned too. Jessica and Donald McDonald were saved."

A sinister theory that was widely discussed at the time was that this was the deliberate action of a rogue Basking shark. This is unlikely but there were incidents following the *Eagle* tragedy which at the time fuelled and supported the rogue shark theory.

"The graveyard near Carradale".

A few days after the Carradale tragedy the *Lady Charlotte*, which was a Campbeltown fishing vessel, was returning home from Arran laden with herring. A member of the crew standing at the stern saw a large shark charge at the boat. The shark struck the propeller a glancing blow. The stern of the boat was lifted 3 feet out of the water by the impact and came down again with a crash, fortunately on an even keel.

Some days later a passenger steamer called the *Dalriada* was coming into Carradale. There was a large number of sharks on the surface which submerged as the steamer came in. One shark however stayed on the surface and made straight for the ship. The animal circled the *Dalriada*, leaping out of the water and lashing its tail. The shark did not attack the ship, but there was a theory held by the locals that this was the same shark responsible for the *Eagle* tragedy. It was claimed that they could identify this shark because it was aggressive and larger than most of the other sharks.

I believe we can safely discount the rogue shark theory and any idea that this was a deliberate 'attack'. I have spent many hours on and in the water off Cornwall watching Baskers and swimming with them. Breaching sharks are a familiar sight to me, and I can readily imagine the scene as a shark came up underneath the *Eagle* and overturned her. I believe the Carradale incident was a tragic long odds accident.

Chapter 2

SHARK SUICIDE BOMBER

The West Briton and Royal Cornwall Gazette dated August 2nd 1956 contains an account of a coroner's inquest at Falmouth. This investigated how two Royal Naval men were severely injured, and two civilians killed, when an attempt to kill a shark by using explosives went tragically wrong.

Lt. Commander J.W. Bailey explained to the coroner Mr L.J. Carlyon what had happened when he was in charge of a team of divers working off Porthkerris Point. His colleague Lt. Commander Brooks was working on *HMS Burly* while the divers were operating from a smaller fishing vessel manned by civilians.

Two divers were under water while a third was swimming on the surface. Brooks saw a shark approaching which he judged to be a dangerous animal rather than the 'harmless' Basking sharks that were commonly encountered. Brooks stated that he had experience of sharks in America, so presumably his judgement that this was a dangerous shark was based on that experience.

He saw the shark swim towards the man on the surface. **The West Briton** newspaper reports 'As it made its run in it turned on its side, which the witness understood was to get its mouth in position to attack. As it came up to the person on the surface, who was unaware of its approach, bubbles came up from one of the divers below. This frightened the shark which broke off its attack and went away'.

The next day at 1.15 p.m. the *Burly* was back in the same place and so was the supporting fishing vessel. *Burly* signalled the fishing vessel that diving should

8

commence and the divers began putting on their gear. As the divers were preparing, Lt. Commander Brooks noticed a shark circling the fishing vessel.

The events of the day before were still fresh in everyone's minds and Brooks sensed there was some anxiety about making the dive.
(N.B. Although I believe it is highly likely that the shark/s concerned were in fact Basking sharks we must remember that Brooks and his men were convinced this was a predator'dangerous to man'. It is not spelt out anywhere but one gets the feeling that they thought the shark was a Great White. The events that followed must be viewed in the context that those on the spot were convinced they were dealing with a dangerous animal.)

There was a dinghy moored astern of the fishing vessel and the civilians Leslie Nye and his friend Richard Kirkby together with Lt. Commander Brooks jumped into the dinghy and set off after the shark with the intention of destroying it. Brooks apparently felt that it was his duty to do this to ensure the safety of his men as he was convinced that the day before the shark had been about to attack the man on the surface.

The account in the West Briton continues...... 'As far as we can tell at this stage they got themselves right up close to the shark which was near the surface and

"The trail started at the offices of Devon & Cornwall media, publishers of the West Briton".

"I finally found the microfiche which told the story of the 1956 shark suicide bomber".

swimming away from them. They had made two charges to straddle the shark, and together Brooks and Spicer threw them at the shark. It was a very good shot. A line linking the two charges straddled the shark and got around either its dorsal fin or its tail. The two charges were hanging either side of the shark with the fuses burning. The boat started to turn away from the shark, but the shark turned around and made for the boat, and was underneath, directly aft, when the two charges exploded'.

(N.B. This is the first time in the newspaper account that we hear of the man named Spicer who was also aboard the dinghy.)

The explosion killed the two civilians Leslie Nye and Richard Kirkby, and the two Royal Naval men, Lt. Commander Brooks and Petty Officer Spicer, were both seriously injured.

The estimated time between the fuses being lit and the explosion was 10-12 seconds, and the explosive was 14 ounces of T.N.T. The court examined whether Brooks had used the right amount of explosive and whether his actions had been appropriate. Lt. Commander Bailey was in the wardroom of the *Burly* having lunch at the time of the explosion. He told the court that he had been told nothing about the dangerous shark being in the area, and thought that diving operations were proceeding as normal. He heard two explosions very close together and thought it was an emergency signal. He went straight on deck where a sailor said "There is a small boat just blown up".'

The small fishing vessel rushed to where the explosion had occurred and was later jointed by *HMS Burly*, which had been at anchor and taken longer to get under way.

As soon as the fishing vessel reached the accident location five men leapt into the water to assist survivors. The bodies were taken aboard the fishing boat which steamed at speed to Porthkerris steps. An eye witness said that the dinghy had been "absolutely smashed to pieces" and he didn't see any sign of the shark.

The coroner commented that it had turned out to be a rather costly experiment. To which Lt. Commander Bailey replied *"The whole question of dealing with sharks is very difficult. I spoke to Brooks about it and he said he was once in an American submarine which got some wires around its screws, and while he and another diver went over the side the Americans held them off with a machine-gun type of weapon"*

The West Briton continued..... ' Witness said he understood sharks sensed when the prey they were going to attack was frightened and it was going to be an easy kill, and that when they smelt blood they lost all fear. On the previous day the man

"*The author and Charles Hood re-enacting the incident*".

on the surface probably looked likely prey whereas the men in diving suits would appear strange. He believed it was accepted among divers that if they released a few bubbles and shouted, sharks would go away.'

The coroner asked Lt. Commander Bailey whether he would have permitted the use of explosives had he known they were going to be used. Bailey responded that he would have wanted to be fully briefed but he trusted Brooks and his judgement one hundred percent.

Another witness was Leading Seaman P.H. Alderton who was one of the naval divers aboard the fishing vessel. He told how he and Leading Seaman Lusty were preparing to dive when someone noticed a shark circling their boat. He confirmed that Brooks had identified the shark as a 'dangerous type', and that everyone was slightly nervous about diving. Alderton further confirmed that it had been Brooks' plan to go after the shark and scare it away with the explosives. Alderton told the court of the double explosion and said he turned to see pieces of the dinghy flying through the air about two hundred yards away from them.

A helicopter was despatched from RNAS Culdrose to Porthkerris with naval Doctor Surg. Lt. A C Johanssen aboard. Leslie Nye's death was due to head injuries and Kirkby died of blood loss. It was stated that Lt. Commander Brooks and

P.O. Spicer were both still very ill in hospital but were progressing well.

And what of the shark? Was this the first ever shark suicide bomber or did the animal survive? Although Commander Brooks felt this was a predatory shark, I suspect it actually was a Basking shark. These are large animals and blowing one up would certainly result in a lot of blood and shark debris at the site of the explosion. The fishing vessel was on the spot in minutes and there are no reports of a live, dead or injured shark being seen.

The Basking sharks I see off Cornwall commonly range between 4 and 6 metres in length. Had this even been a Great White, Mako, or Porbeagle shark of say 2-3 metres, it is still strange there was no evidence if the shark was blown up.

The incident was big news in the area and the explosion occurred quite close to land. So had shark parts been washed up there would certainly have been reports – I have looked and found nothing.

This incident like the one in Carradale some twenty years earlier, was a tragic accident. Unlike Carradale, however, in this case man certainly had a hand in his own destruction.

Chapter 3

TWO CHEFS

If you are in the restaurant business and your name is Smith then statistically as far as shark attacks go you are in a very high-risk group!

We've looked at the tragic accident at Carradale and also at the bizarre disaster of the attempt to blow up a shark. Our shark attack trail now traces two really quirky tales of restaurant workers called Smith both being attacked by sharks on land!

The Sunday Times dated 9 September 2001 told the story of **Darren Smith, a chef from Newquay, Cornwall**.

Darren was driving a seven-foot shark to a restaurant. The shark was on a bed of ice in the back of his van. He braked sharply and the 110lb shark shot forward and ended up with its mouth on his shoulder. He reached over to grab the shark's nose and push it back. It seems he missed the nose and ended up with his hand in the shark's mouth. Whether the shark's mouth closed on his wrist, or whether he just caught his wrist on the shark's teeth I don't know. However he suffered a severed artery which would have made the inside of his van a very bloody place to be sitting.

"The nurses at the hospital couldn't stop laughing" he said. "I must be the first person in history to be attacked by a shark on dry land!"

As we will see he wasn't the first and probably won't be the last. Being on dry land in Britain is no guarantee of being safe from shark attacks, in fact statistics

"Darren braked, and the shark shot forward onto his shoulder".

show you are safer in the water! Read on and you'll find out that you are not necessarily safe from shark attacks even in the toilet!

Recreating the Darren Smith in-vehicle attack for the film posed problems. Just where do you find a large predatory shark in England that you can take for a ride in a van? How do you get a dead shark to perform on cue and leap forward to attack its victim? This dead actor has to be prepared to do several takes until the scene has been filmed to the director's satisfaction. Enter Mable, a 7 foot shark caught off Looe by the Shark Angling Club and given to a taxidermist to make immortal. The club gave her to me in a rather dilapidated condition in 2007, but after a hospital visit to my sculptor and modeller friend Scott Gleed, she was restored to her former glory.

Mable isn't any ordinary stuffed dead shark, in fact she is a bit of a media megastar in her own right. She helped me launch my book *Sharks in British Seas* when she was seen touring north Cornish towns relaxing on the roof of my Subaru. At the same time she appeared in various newspapers and magazines, and has been in front of the camera on many occasions.

Mable rose 'girlfully' to the challenge and was taken in a car from her home by the sea in north Cornwall across the peninsula to the Shark Bay film studios in

"An unusual sight to see in your rear view mirror!"

Porthleven, south Cornwall. *(N.B. She doesn't really like riding inside cars and far prefers being on the roof of vehicles in the fresh air. She told me this will be a stipulation in her next film contract!)*

She had to be helped and pushed forward to make her in-car attack on the actor playing Darren, but generally her performance was perfect and is a testament to her acting skills. She drew blood when I got her home – my blood – I caught my finger on one of her teeth when my wife Jacqui and I were putting her back in her cradle in her beach house. Was it deliberate? I'll never know, but I fancy I detected a twinkle in her eye as she watched me suck my finer. Perhaps her acting fee wasn't enough!

— — — — — —

Our second chef shark attack victim was Paul Smith. The Blacktip Reef shark's usual home is tropical reefs. A landlocked pub in rural Worcestershire would not be where you'd expect to find sharks, and certainly not where you'd expect a shark attack. Miami, the Blacktip, measured 2½ feet and his British residence was a 3,000 gallon aquarium in the Fountain Inn pub near Tenbury Wells. The young Blacktip shared his home with 1½ foot catsharks Forest and Jenny.

"*The pub shark attack made headlines all over Britain*".

"Richard interviewing chef Paul Smith".

Paul was feeding Miami fish bits while his diners watched. Paul wasn't quick enough and the Blacktip grabbed his fingers giving him wounds which required stitching in the nearby hospital. The sharks had smelt the prawns and Paul's blood, and became very excited.

Paul told his story to the Sun newspaper. "They must have got a whiff of the prawns – they shot across the tank like bullets. Within a second the water was churning like the shark attack scenes in the Jaws movie – and my hand was in the middle of it. Miami did the most damage. He was hanging off my finger and I was howling. The nurses thought I was mad when I turned up at the hospital and told them a shark had got me".

Paul's boss, restaurant owner Russell Allen who was a shark fanatic and had installed the 15 foot tank, rushed Paul to hospital. Russell told the Sun, "Diners love watching the sharks at feeding time, it must help them work up an appetite – seafood sales have gone up 40 percent". He added, "Paul was very lucky to have only needed a few stitches".

I interviewed Paul for the film *Shark Attack Britain* and here is his account of the incident as it appears in the film.

Paul - *That evening I took the prawns from the fridge, dropped them into the tank, and took my eye off them for one split second. There was blood and a few screams and before I knew it I was on my way to the local hospital.*
Richard – *Took your eye off them for one split second and then what? He attacked you, he bit you, he got your arm, what?*
Paul – *Before I knew it he was on my finger.*
Richard – *Have you got any scars or anything?*

Paul – *Yes I have there is one on that finger there.*
Richard – *3 ½ inches worth of wound. Was it the first shark attack wound the hospital had dealt with?*
Paul – *I think so, the first one on record.*
Richard – *So you are the first victim to be attacked by a shark. How many miles from the sea are you here?*
Paul – *Roughly 100 miles give or take a few miles.*
Richard – *That's pretty bizarre, you get attacked by a shark in a pub a hundred miles from the sea.*
Paul – *Yes, it was totally bizarre.*
Richard – *How big a monster attacked you?*
Paul – *He was nearly 3 feet.*
Richard – *And do you think this shark preferred prawns or chef?*
Paul – *Definitely prawns.*

Thus far we've had an accidental boat overturning in Scotland which resulted in three people drowning, a Basking shark blowing up the Navy, a dead shark attacking a chef and a live shark biting another chef in a pub. It's beginning to look as if swimming in Britain's seas is pretty safe, compared to being on land and in boats!

Let's see...............

Chapter 4

(pub & toilet, you're not safe anywhere)

The attacks we have looked at so far have involved whole sharks, both alive and dead. *Shark Attack Britain* moved into even more weird and bizarre waters when I started discovering that people had been injured by jaws that were no longer attached to their sharks. Two blonde women living less than fifty miles apart were both 'attacked' by jaws they keep as ornaments.

The Wheatsheaf Inn at Bough Beech was once a hunting lodge and dates back to the time of Henry VIII. The walls of both bars in the pub are crowded with the heads of animals which were hunted in bygone times.

Also on the wall is a heavy block of wood on which is mounted a shark's jaw. This interesting object has its own story and is called 'Old Gory'. In 1826 a shark was reported to have taken three sailors from the vessel the *Java Queen*. The shark was killed and the dead men's shipmates mounted the jaw on a block of wood framed by an old piece of rope from the *Java Queen*.

This grisly relic was discovered in a Westcountry antique shop by Liz Currie, the landlady of the Wheatsheaf Inn. In 1997 'Old Gory' was disturbed and fell off the wall and onto Liz's head and then her shoulder with the sharks teeth puncturing first her scalp then her neck and shoulder. I felt the weight of 'Old Gory' and it was considerable, which makes Liz Currie's escape without serious injury remarkable.

"Liz Currie and her 'attacker', Old Gory".

This is Liz's story as she told it in the film.

"In the summer of 1997 on a hot July afternoon Old Gory was mounted in the centre of our menu board. I stood on a chair and leaned in front of him. Unfortunately I lifted my shoulder which caught underneath his jaw bone. He fell onto me with his top set of teeth sticking into my head and the bottom very very sharp needle teeth scoring through my shoulder and a very very nice silk shirt that I had, causing a lot of blood and a lot of shouting….. and a lot of upset amongst the customers who were obviously deeply concerned and shocked to see this shark attack actually happening in their local pub."

As can be seen in the photographs the shark's teeth are very sharp like needles. With the weight of the wood helping them to dig into first Liz's head, and then her shoulder, the resulting ripped blouse, some puncture wounds, and a lot of blood loss all meant she had to go home and change her shirt, dress her wounds and clean herself up. Remarkably she came straight back to her pub and resumed pulling pints!

The media got hold of the 'pub shark attack' and the story went around the world. The usual red top tabloids (the Sun, Mirror and Star) all loved it as did the Mail, the Express and others. The story even made the Australian Womans Weekly!

"The Wheatsheaf Inn, Kent. Home to Old Gory and owned by Liz Currie".

— — — — — —

In July 2009 Tess Bath and her boyfriend Tom moved into their new flat in Tulse Hill, south London. One of Tess's most treasured possessions is a sharks jaw which moved in with them. Tess is a shark fanatic, and she was given the jaw by a family friend whose great grandfather had caught the shark.

Tom watched with bemused tolerance while Tess paced the flat carrying the jaw, searching for the best place to put it. Lots of locations were tried but the place it seemed most at home was behind the lavatory sitting on the cistern. This position meant that anyone flushing the loo had to put their hand behind them through the shark's jaw to press the flush.

Just over a week after the jaw had been in position Tess went to press the flush and the jaw slipped and crashed down onto her wrist. Sharks teeth are sharp and she soon found out how sharp when they broke her skin in several places, drawing blood. Plasters and a bandage were required and the scars took nearly a month to heal.

The jaw has since been inactive and Tess and Tom have been safe – but for how long? Following the accident the jaw has been tried in several places and currently resides in the kitchen by the phone. Soon after the attack Tess noticed the jaw had a tooth missing – was the tooth lost when the jaw hit Tess' arm? We'll never know, and the tooth has never been found.

Here's the story in her words.

"I had been given the most wonderful rather large Tiger shark's jaw by a family friend. I had the dilemma when we moved house of having to decide where to put it. Not having anywhere appropriate anywhere else in the house, I thought perhaps on top of the cistern in the loo might be a rather nice idea, so that when someone went to flush, they would have to reach their hand through the jaw itself. So of course I daringly was the first person to trial the hand-through-jaw flush mechanism. As soon as I touched the top of the cistern, the whole thing rocked and fell down and bit me rather convincingly on the wrist. It was horrendous, torrents of blood. I was screaming. No, it gave me a nasty old bite, I did have a lot of puncture marks. I don't know where the shark was caught, but it has still got it!"

"Tess Bath and her partner Tom with the jaw that bit her in the toilet".

Chapter 5

UNDER A NORTH SEA OIL PLATFORM

THE TARTAN OILFIELD IN THE NORTH SEA, JUNE 1988

One hundred and fifteen metres (370 feet) below the surface a diver is doing a video inspection.

From Diver magazine
"ACU Jimmy diver was doing a video survey around the slug-catcher on Tartan. (Don't ask what a slug-catcher is). He's in 115 metres of water, he's relaxed and happy in his job, using the handheld video camera. Up in the dive shack normality fills the air. Suddenly, his voice bursts through on the comms.
DIVER: *"Oh, my God, look at this bastard." He pans the camera to reveal a ten foot Mako shark swimming close alongside, it's big eye-ball staring menacingly into the lens. The supervisor was obviously born with the innate ability to state the obvious.*
SUPERVISOR: *"Bloody hell, that's a shark!"*
DIVER: *"Yes I know". The diver stops, keeping the camera trained on the slowly moving shark. The Mako then swims quickly off into the distance and out of sight. The stunned diver's limited concentration is shattered by the disparaging words of the supervisor.*
SUPERVISOR: *"Is it bothering you?" This question was obviously misinterpreted by the diver who thought "bothering" was a marine biologist's term for attacking or eating.*

"North Sea oil rig typical of those found in the Tartan oil field".

DIVER: *"Not at the moment".*
On the screen there is no sign of any shark, and the diver returns to his survey. We can assume that his mind was not completely on the job in hand. It was while the diver was attempting to continue with his work that the shark made its move. Suddenly, the camera is turned to reveal the shark moving in, its body rolling over as it comes in for the snap. Teeth and white belly flash across the screen, the picture blurrs and then goes haywire; it is the realized image of every diver's worst nightmare. With no protection to defend himself this brave C.U. Jimmy diver did the only thing he could in the circumstances, he rammed the camera down the shark's throat.
SUPERVISOR: *"Diver, diver, are you alright?"*
DIVER: *"Ooooo! Ooooo! Ooooo!" The air is filled with the sound of panic breathing.*
SUPERVISOR: *"Diver, diver, are you alright?"*
DIVER: *"Ooooo! Ooooo! Ooooo!"*

For over a minute the supervisor is unable to get any sense out of the diver. The shark has swum off, and from the noise coming from the comms its clear the diver is none the worse after his ordeal. When calm is restored, the supervisor passes on some unwelcomed news.
SUPERVISOR: *"We've got it all on video".*
DIVER: *"I know you f***ing have, I hit the bastard with it. Oh, sod this, I'm going back to the bell."*

"Texaco operated the Tartan oil field".

"Porbeagle sharks are often sighted around North Sea oil rigs".

It is more likely that the shark was a Porbeagle than a Mako. What was really annoying when making the film *Shark Attack Britain* was that we couldn't find the actual footage taken on the diver's camera. I knew it existed because I'd seen it. The shark appears out of the gloom and swims straight at the camera, there is an obvious impact – the diver hitting the shark – and the shot goes haywire.

I don't think the shark was attacking the diver, I believe it was probably attracted by the light and moved in to check it out – then got a hefty clout. An attempted attack or not, it must have been terrifying for the diver to be charged by a shark, and when all you've got in your hands to fend off the charge is a camera, that's what you have to use.

The shark was never seen again and the video of the encounter was hugely in demand as far as the north sea oil community were concerned.

Chapter 6

ODD BITES

Mid September 1785 – The Brighton shark

IIn September 1785 an anonymous letter was sent to the London Morning Herald which described an incident which became known as 'The Brighton Shark', and which was still being talked about a quarter of a century later.

It was unusually warm for September and large numbers of people were paddling on the beach, and swimming in the water. One swimmer was some distance out when he saw a shark fin coming at him very fast. The man swam to shore as quickly as he could and was scrambling up the beach when the persuing shark launched itself at him, and stranded itself on the beach. An eye witness described the incident as follows.

"A number of people, with hatchets, attacked the fearsome creature and killed it. On opening its stomach, the entire head of a man was found inside it, not otherwise altered than being very soft and pappy….The shark was twelve feet in length from its head to its tail".

The shark was named as a Tiger shark and news of this 'attack' caused widespread alarm and kept people out of the sea.

Note. Given that this report is based on an anonymous letter to a newspaper, and there were no corroborating witnesses or stories, it seems possible that the writer was a hoaxer, or a serious Brighton beer drinker! The only other report of a

Tiger shark attack in our waters was of a diver being threatened in July 1968 off Cornwall.

July 1812 – Mill Bay, Devon

It was reported that an unnamed soldier from the Lancashire Militia was swimming in Mill Bay in July 1812 and was attacked and severely wounded in both legs by a shark. The shark was said to have been caught and killed a few days later by local fishermen. Apparently the shark had been eating mackerel and this knowledge suggests that the fishermen cut it open to inspect its stomach contents. **If** the incident is true the shark is likely to have been a Porbeagle.

July 1848 – Hunstanton, Norfolk

"A man was standing waist deep in the sea when he saw something of a most formidable size approaching. He hit out at the creature as he was alarmed, but the animal fought back and struck him with his tail. The animal was a nine foot long shark and a struggle followed which left the shark stranded. Once stranded the shark was killed, apparently the same animal had been behaving aggressively towards bathers earlier that day!"

September 1864 – Granton, Edinburgh

It was reported that a man called Ballard was bitten three times on the leg by a three foot shark. Both man and shark survived, and the shark was further reported to have been seen later in the day by other swimmers.

1876

In his book *Shark Attacks* Alex MacCormick tells of an incident which occurred in 1876 between Hastings and Fairlight in Sussex. A bather had swum the 400 yards to an anchored smack, and on his return swim found himself carried by a strong ebb tide. He felt something come into contact with his left leg and struck at it. He felt his hand rub along a large fish. He yelled out and swam away as fast as he could, but the fish scraped along him another two or three times. He attracted the attention of a nearby fishing boat which came to his rescue. Apparently he almost 'vaulted aboard' in his eagerness to get out of the water.

A large fish was seen swimming around the boat which the report identified as a Blue or Porbeagle shark.

The Times, London, 14 August 1919

DEVON, UNITED KINGDOM, 1919
"A shark which had made its appearance among women bathers at Croyde, North Devon, on Tuesday was the cause of considerable excitement. The bathers got safely ashore, and the shark was shot by Mr C.C. Cuff, assistant manager in the Great Western locomotive works, Swindon. It took five persons to drag ashore the shark, which measured 7ft 6in in length".

The Times, London 15 July 1924

The paper relates that a twelve-foot shark (unidentified) was caught by fishermen who were netting mackerel. The shark broke two men's arms before it was hauled into the boat.

I would hope that today fishermen would merely cut the shark free and release it thereby not risking either themselves or the shark.

Kent Coast, July 1971

In his book *Shark Fishing In British Waters* Trevor Housby tells of an 'attack' off the

Kent coast when two large Thresher sharks were said to have attacked a child swimming in very shallow water. Housby states, "I am inclined to think that this attack was more of a chance meeting than a deliberate attempt at man eating". **I agree**.

The People, London 16 April 1995

"A mother and her three children were treated by ambulance men after inhaling fumes from a dead shark. The one-foot fish, kept in formaldehyde and other toxic chemicals, had been left at their home in Brentford, London".

The Isle of Wight, September 1995

An article in Diver magazine reported a shark encounter 6 miles off the Isle of Wight.

"Wreck diver George Hayward was diving the wreck of the *SS Westville* with his buddy Trevor Jones. The *Westville* lies in 40 metres of water. Hayward had an unexpected meeting at around 20 metres. "This shark, of about two metres, brushed against my mask as it went by, I was startled as it passed right against my nose".

No attack, so the pair continued their dive and didn't see the shark again. The diver thought the shark was a Porbeagle which, although common throughout British waters, is rarely sighted by divers. Times have changed: In 1995 this was probably an incident divers would have preferred to avoid, now 15 years later many divers are desperate to get into the water and get close to sharks.

Northern Scotland, June 27 1960

In his book Shark Attack Mac McDiamid records how a young German fisherman, Hans Joachin Schaper was bitten on the arm by a small shark he was removing from a net. Hardly an attack, and such incidents are common, nevertheless this is worth mentioning because the incident is listed as an 'attack' in the International Shark Attack File.

Schaper was aboard the *Mai* and ten days later his wound had gone septic. He was transferred to another East German trawler the *Karl Marx* which landed him at Wick harbour so that he could be given hospital treatment.

Was he really attacked by the little shark or was this all a communist ploy to put a man on the British mainland into one of our hospitals so that he could spy and steal our medical secrets?! We will never know, but what we do know is that the word 'attack' conjures up all sorts of images, and in British waters the images are almost always the wrong ones.

Chapter 7

TOO CLOSE FOR COMFORT

An incident occurred off the Isle of Wight in June 1981 which could easily have added to the (thankfully) small tally of human deaths caused by sharks in British waters.

Danny Vokins is a well known shark angler living on the Isle of Wight. On the day of the incident he was out shark fishing on his friend Ross Staplehurst's boat.

They had been chumming all morning, with baited hooks and lines out. They knew there was a shark around. In Danny's words he could almost 'feel' it, and it felt as if the shark was stalking them.

Suddenly, just after 2.00 p.m., a 181 kilo (398 lb) Thresher breached right beside the boat and then a second later landed in the boat with the (more than a little) surprised anglers. The shark was too big and heavy for those aboard to be able to lift back over the side and into the water. It was with great sadness that they had to kill the shark, but it could have been the safest thing to do because the tail on a Thresher that size could kill or seriously injure a human. Hazardous though the tail could have been there is also the little matter of the miracle that no humans were injured when a 28 stone shark landed among them.

"Danny and Ross commented as follows.
Danny: *"We were out fishing a club competition and after about 2 hours we saw a big shark jump about 1/2 a mile from us. 3 or 4 minutes later it jumped again, this time only 300-400 yards away, another couple of minutes went by and it*

"The author in the office talking to camera describing the IOW incident".

"They had a lucky escape, that tail could take your head off".

jumped right alongside the boat. It got caught in a fishing line and took off at a rate of knots. Then it jumped again about fifty metres away and a split second later it jumped into the boat. It had been getting closer all the time, almost like it was stalking us. The story got into the press, and went worldwide. We were 5 people in a 25 ft boat with a 14 ft 400 lb shark. Its amazing no-one was hurt or even killed".
Ross: *"The boat lurched right over. It hit the boat so hard it either killed itself or knocked itself out. If we could have put it back in the water we would have, but it was too big to lift and a fish that size could have been very dangerous so we had to kill it for safety's sake."*

Thresher sharks are not uncommon around the Isle of Wight in the summer months and 15-20 sharks are caught, tagged and released in this area each year.

On 21st November 2007 the largest Thresher shark ever landed anywhere in the world was caught off Lands End, Cornwall. This giant shark weighed 510 kgs (1122 lb, 80 stone) and measured 5 metres in the body and just under 10 metres if the length of the tail is added. The shark was a female, and she probably could have been released, but it was rumoured that the fisherman brought her in because he hoped to make money out of the media interest that would certainly occur. According to the rumour mill he didn't make much money, and in fact he got less than it cost to repair the damage to his boat and nets that the shark had caused.

This fantastic breeding-age specimen was sold at auction for only £200 and apparently ended up in a landfill site – tragic.

Danny Vokins and Ross Staplehurst are still Thresher shark angling around the Isle of Wight and all the sharks they catch are released.

Like Porbeagles, Makos and Great Whites, Threshers are semi-warm blooded and maintain their body temperature at a constant level. Film has recently confirmed what researchers have long suspected which is that Threshers use their formidable tail as a means of stunning and gathering up their prey.

Chapter 8

BEESANDS MAKO

Jim Johnson was a member of the Aldershot Dolphins diving club. It was the club's custom to take an annual diving holiday, and in 1971 the site they chose was Beesands in South Devon which they visited in June. The incident described in this chapter took place on June 12th.

After unpacking in his caravan Jim decided to go for a snorkel and took with him his 'lobby' hook in case he saw a crab or a lobster which he could take for supper. He swam out and was about 50 yards from shore in 25-30 feet of water when he found he had company. **In his words as reported in Diver (Triton) magazine:** *"All of a sudden there he was! A shark. My shark. The only shark in my life. He looked as big as a bus – a 12 foot long bus that is. Allowing for water magnification he must have been a good 8-10 feet long.*

His first thought was that it was a Basking shark, his second thought was that Jacques Cousteau's method of repelling sharks was to bang them on the snout with a wooden pole. His thinking got more serious as he realised that this wasn't a Basking shark but perhaps a dangerous shark."

In his own words Jim was terrified and genuinely believed he was in a situation which threatened his life.

"The shark had swum a little ahead of me, his own head about a foot below the surface. He suddenly turned and came at me fast head-on. Although I had a knife strapped to my leg I had no thoughts of getting it out. I just stuck my 'lobby' hook

"On the beach describing the Beesands incident".

"We'll never know the identity of the Beesands shark - Porbeagle, Mako, or even a Blue shark".

out, pointed end first.

The hook hit him where you would think his forehead was and he passed closely underneath me and then circled me about five feet away. I could see every detail, his snout, his hungry looking mouth, his tough abrasive skin. He also had a silver coloured remora just below and behind his gill slits.

After what seemed an eternity of circling around me he flicked his tail and shot off into the blue. Panic took complete control of me. I swam madly for the shore with my head down in the water looking out for him. I must have made more commotion than a drowning man or a dying fish for he came back and took up his awful circling again. I just kept finning and hoped for the best.

Luckily, I was soon in 10-15 feet of water, and he left me once again. I crawled up on to the beach and collapsed in a state of exhaustion, emotional rather than physical.

A friend, Brian Poulter, came down to me. "Find anything" he asked. When I gabbled out my story he thought I was kidding, but just then a second club member, John Timpany, came running down to us. "Did you see that shark" he said. Did I see that shark! After I told him what happened he said, "I thought you must have seen it because he was so close to you". Quite a comedian is John.

After a drink to calm myself I tried to identify the shark with the various fish books owned by club members. As far as I could tell it was either a Porbeagle or a Mako. I go for the Mako because it is the slimmer of the two, and from what I could remember my shark was slim – like a torpedo.

Why didn't he take a lump out of me? I don't know. Apparently so many factors could influence him one way or the other. I was told later by the locals that seals were common the other side of Start Point, so he may have thought I was a seal.

Probably what deterred him from taking any further action was the low water temperature or the smell of my neoprene wet-suit (or me). I like to think that what really stopped him was the thump he got on the head with my 'lobby' hook. Anyway that's my story and I'm sticking to it.

Although a shark was sighted by club members from our boat on two occasions later on during the holiday, he was never again see by divers actually in the water.

There was one poor chap, though, who was given the fright of his life by a black polythene bag at 30 ft!"

"John Timpany being interviewed on Beesands beach".

The aforegoing section in italics has been reprinted from Diver magazine (then called Triton) and is the copyright of Diver magazine – thank you to Diver.

Porbeagle or Mako? We'll never know but the likelihood is that it was a Porbeagle simply because they are commonly found in our waters while unfortunately Makos are rare.

Jim's fear was totally understandable even though it is more likely that the shark was being curious than attempting to attack him. This incident occurred nearly forty years ago, and in that time our knowledge of sharks and our attitude to them and the threat they pose has changed dramatically.

We interviewed Jim's friend John Timpany for the film and here is what he told us.

"We arrived down here on holiday with my family. It was about late afternoon in June and the weather was beautiful, lovely sunshine, glassy sea, flat calm, and he was snorkelling out there, and I was watching him as I was eating, and then suddenly he was only about 60/70 yards from the beach, and I saw this fin go past. I said to everyone else in the caravan did you see that shark, and they just looked up and said no. No, didn't see it, you must have been imagining it. I was looking again while I was eating and it came back again, a fin went past, it was only

visible for a few seconds, but it went past him again from a different direction. And again I said did you see the shark, did you see the shark, and they looked up and said no. By that time Jim Johnson was snorkelling in and we got out of the caravan and walked down to the beach to see if he'd caught anything, and then as soon as he got to the shallow bit and he could stand up, he was blurting out that this shark had molested him and it must have been about 8 to 10 feet long, and it came up so close that it would bump into him and he would push it away with his hand spear. Obviously he got quite flustered, snorkelling back as quickly as he could, and then again it came in from the other direction and he pushed it away again. He was really quite frightened and he was explaining all this to us on the beach, which clarified to the others the fact that I saw it."

Chapter 9

LUNDY BLUES

The shark that angler Stephen Perkins caught near Lundy Island at the end of August 2008 first took his baited hook and then his arm. Fifty two year old Stephen from Penarth in south Wales was angling with three friends aboard their 24 ft. boat when he caught the Blue shark As he was preparing to unhook the shark prior to its release, it sank its teeth into his arm leaving him bleeding badly, and eventually having to have reconstructive surgery.

The shark was released but Stephen was clearly in a lot of trouble. The bite had severed arteries and the blood loss had to be stopped. Stephen's friend Andy Samuels bandaged the wound, applied a tourniquet, and made sure that Stephen kept his arm in an elevated position. The dressing was soon soaked through, and Andy realised that if the bleeding couldn't be stopped his friend would pass out and eventually die.

He got on to his mobile phone and called the Swansea coastguard. Despite its being a bank holiday weekend the response was instant. Luckily for Stephen an RAF Sea King rescue helicopter was exercising close by and was hovering over their boat in a matter of minutes. Stephen was winched up to the helicopter and whisked to the emergency unit at the North Devon Hospital in Barnstaple.

Stephen was quoted as saying, "The scariest bit to be honest was being winched up to the helicopter. It won't put me off fishing again, but next time I will remember to pick the shark up by the blunt end. I've had a few scrapes in the past but this is the first time I've been bitten by a mini version of Jaws".

The 'victim' Stephen Perkins being interviewed with Andy Samuels whose quick thinking saved his life".

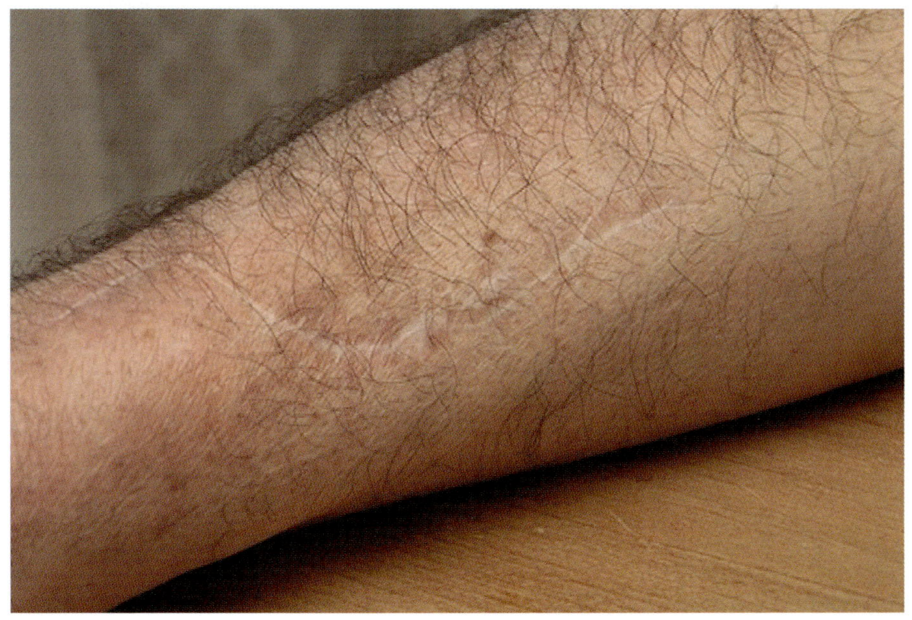

"With severed arteries Stephen could easily have bled to death".

"A photogenic Blue shark poses for the camera".

The quick thinking of Stephen's friend Andy saved him from possibly becoming the first person in Britain to die as a result of a shark bite. The interview from the film went as follows.

Andy: *"There were four of us on the boat there, three of us have caught sharks before. We got the shark to the side of the boat, we went to take a photograph, and the shark spun up in the trace when he was alongside the boat, with his head sticking up. Steve tried to grab him so he could cut the trace and he spun around a bit too quick and grabbed hold of Steve's forearm.*
Steve: *"I turned around to talk to one of the boys, lost my concentration and the shark latched on to my arm, and his bottom jaw was in the flesh and the top jaw stopped on the bone there, but it had severed two arteries, and the bone was showing.*

Steven's wound was potentially life threatening because they could not stop the blood flow.

Richard: *"If you had not been winched off by helicopter and flown to hospital, how long were you from passing out – an hour?*
Steve: *"Yeh, I would have bled out, my blood pressure would have dropped, and I would have been in deep trouble."*

Blue sharks have been a target of British anglers for over 60 years. In the early 1960's the fleet of the Shark Angling Club based in Looe, south Cornwall, recorded annual Blue shark angling catches of over 2,000 sharks. In one year the figure was over 6,000, since then angling catches have plummeted and last year (2009) only 149 sharks were caught by the Looe club.

Stephen Perkins and his friends always release all their sharks as do the Shark Angling Club and most other anglers in Britain.

Chapter 10

MASS ATTACK
SHARKOCIDE

Sharks in British Seas that might be considered dangerous to humans in some circumstances.

Basking sharks *(Cetorhinus maximus)*
Blue shark *(Prionace glauca)*
Porbeagle shark *(Lamna nasus)*
Shortfin Mako *(Isurus oxyorinchus)*
Thresher shark *(Alopias vulpinus)*
Smooth Hammerhead *(Sphyrna zygaena)*
Great White shark (if present) *(Carcharias carcharadon)*

Undoubtedly the most dangerous animal in/on the sea

Humans *(Homo sapiens)*

– – – – – – –

The shark attacks we have looked at so far have ranged from the tragic to the bizarre, the comic, the weird and the ridiculous, and they all share one thing – sharks were 'attacking' humans.

The reality of 'Shark attack Britain' is that sharks have much more to fear from man than man has from sharks".

If we turn the coin over and look at man attacking sharks a history emerges of massive attacks on several species. I often think that one of the biggest problems that sharks have is that they are called fish. Because they are fish they are seen as a food source and are fair game for harvesting. Because some species very occasionally injure or kill humans, sharks don't usually get a sympathy vote like dolphins and other cetaceans.

In many ways some sharks are almost more like mammals than fish, and some species reproduce similarly to mammals with low reproductive rates. This has not saved sharks from being killed in huge and unsustainable numbers – they are fish. The insatiable demand for shark fins for soup in southeast Asia has led to sharks being killed in increasing numbers in recent years. Some shark species now have protection in British and northern European waters, but this wasn't always the case, and over the years our waters have seen the brutal slaughter of several species in large numbers. **This is the real 'Shark Attack Britain' story.**

The reader may have smiled at the 'attacks' in the pub, lavatory, van and restaurant. There is nothing remotely funny about the 'attacks' we will look at now. Basking shark fisheries existed in our islands up until the 1950's. This huge fish weighing up to six to seven tonnes was hunted chiefly for its liver oil which was used as a lubricant, in lamps, and the squalene extracted from the oil went into cosmetics. Its skin has served as sandpaper and for making leather goods, while insulin was extracted from the pancreas. Basking shark flesh has been converted into fish meal and of course its fins end up in soup. Lots of reasons to kill sharks, but perhaps the most bizarre was during the Second World War when Hurricane fighter pilots training in Scotland sometimes used these gentle giants for target practice!

Gavin Maxwell, the author of *Ring of Bright Water*, and Anthony Watkins both operated Basking shark catching operations after the Second World War and into the 1950's. Sharks were harpooned like whales and dragged back to factory sites ashore for butchering and processing. It is possible that up to 100,000 Basking sharks were killed in the northeast Atlantic, including the British Isles, during the last century.

In December 2006 the Angel shark was declared locally extinct in the North Sea by ICES (the International Council for the Explanation of the Seas). Extinct in the North Sea and severely depleted around the rest of the British Isles, the Angel shark is the first of our sharks to be declared locally extinct.

The Spurdog shark or Spiny dogfish will be familiar to many readers under another name – rock salmon. Sold in fish and chip shops up and down Britain this

"Sharkocide - over 130 Porbeagles were killed in one place in one week".

48

"Fish don't feel pain!? - this shark certainly isn't having a lot of fun".

species is thought to be depleted by up to 95%.

In 1961, 6286 Blue sharks were caught on rod and line by the shark angling fleet operating out of Looe in south Cornwall. In 2009 the Looe shark anglers recorded catching 149. This is a relevant and very local British example of a population crash. It's not the shark anglers who have caused the damage, it's the massive over-fishing of these sharks by commercial fisheries, notably longliners. To see this wonderful, sleek, beautiful electric blue shark in the wild is a privilege. To see large numbers of twisted and lifeless bodies lying on cold floors in markets, having changed colour, is a tragedy.

Our film *Shark Attack Britain* ends with footage of Cornish fisherman Martin Ellis catching Porbeagle sharks off Newlyn in December 2003. It is gruesome footage accurately depicting a horrifying episode in which over 130 Porbeagles were caught and killed. In July 2007 an Appledore-based longliner caught 60-90 Por-beagle sharks in a single day near Lundy Island. Like the Newlyn incident this made national news, but things were starting to change because most newspapers rightly condemned what had happened.

Porbeagles, Basking sharks, Angel sharks and Spurdogs all now enjoy varying measures of protection in British and northern European waters. But hundreds of thousands, indeed millions of Blue sharks, Makos, Threshers, and Hammerheads are still being killed in the north Atlantic each year. For them it's still shark attack open season.

Shark Attack Britain *– who attacks who?*

who is to blame?

who is guilty or innocent?

who eats who?

I think you now know the answers.

"The author at Newlyn".

THANKS AND ACKNOWLEDGEMENTS

Tess Bath

The Carradale Hotel

Dr. Paul Chambers

Liz Currie

Tim Davison

Dennis Fairfax – author,
The Basking Shark in Scotland.

Denise Headon – secretary,
worn-out typing fingers

Diver Magazine

Trevor Housby – author,
Shark Fishing in British Waters

Mable - *stuffed Blue shark*

Alex MacCormick – author,
Shark Attacks

Mac McDiarmid – author,
Shark Attack

John Nightingale

Chrystal Patterson

Jacqueline Peirce – wife,
vital support

Steve Perkins

John Boyle – Shark Bay Films

Fionn Crow Howieson – Shark Bay
Films

Eddie Milano Rourke – Shark Bay
Films

Andy Samuels

Simon Spear

The Shark Conservation Society

The Shark Trust

Paul Smith

John Timpany

Jed Trewin

Danny Vokins

Special thanks to Eddie 'Camera Eye
Rock n' Roll' Milano Rourke

About the Author

I have been lucky enough to have had a long and glittering career with more ups and downs than a 90-year-old getting the birthday bumps. It has certainly had as many downs as ups! My education in Cornwall and then Devon came to an abrupt end when I left school in the middle of a sixth form term to join a rock and roll band and become a megastar. However, as starvation was more of a possibility than stardom, I decided to move into managing rock bands and capped this part of my career by making the 'brilliant' decision to turn down a singer who went on to have a huge hit that for many years held the record of being the longest running British Number One. In fact, so good was my rock and roll judgement that, together with Paul Gardner and Terry Sullivan of Rainbow Reflection, I also made the 'clever' decision not to follow up on an opportunity to work with David Bowie – a year later he was a superstar. Our time in horseracing was marked by a similar discerning decision. My wife and I decided we couldn't afford to buy a yearling colt for £5,000 which went on to earn over £1,000,000 for its owner.

The Peirces have a long 'East of Suez' history. My great-grandfather started Peirce Leslie & Co. in 1862, which became one of South India's leading companies. My uncle and cousin followed in his footsteps and also had distinguished careers in the Subcontinent. After spending time as a 'guest' of the Japanese on the River Kwai in World War Two, my father spent most of his Army career in the Middle East, where I too, have spent a large part of my life.

I have been fascinated by sharks ever since I was told I couldn't go swimming because there had been a shark attack. I was a child in Kuwait then, and now here in Britain sharks have become my life. I am chairman of the Shark Trust and the Shark Conservation Society and feel privileged to have met and worked with these beautiful creatures all over the world.

© *Jed Trewin*

© *John Nightingale*

Shark Attack Britain – *who attacks who, who is to blame, who is guilty or innocent, who eats who? I think you now know the answers.*

Useful Shark Contacts

American Elasmobranch Society ....www.elasmo.org

ElasmoFrance...www.fredshark.net

Elasmo.com ....www.elasmo.com

Fishbase ....www.fishbase.org

International Shark Attack File......................www.flmnh.ufl.edu/fish.ISAF/ISAF.htm

Italian Research Group............................http://digilander.liberto.it/infogris/

Mediterranean Shark Site www.zoo.co.uk/-z9015043

Reef Quest Centre for Shark Research............................www.elasmo-research.org

Richard Peirce ....www.peirceshark.com

Shark Alliance ....www.sharkalliance.org

Shark Conservation Society...www.sharkconsoc.com

Shark Cornwall..www.pierceshark.com

Shark Research Institute ....www.sharks.org

Shark Trust..www.sharktrust.org

www.sharkconsoc.com